Where it Began...
The Impetus for Transformation

On December 9, 2010, the Telework Enhancement Act on 2010 was signed and is now Public Law 111-292. This Law now paves the way for Federal Agencies to design telework programs that improve COOP (Continuity of Operations), promote management effectiveness, and enhance work-life balance.

As part of this Law, Agencies were required to complete the following within 180 days, or by June 2011:

- Establish a policy authorizing eligible employees to telework.

- Determine the eligibility of all employees *of the agency to participate in telework* and *notify employees* of their eligibility status.

- Require a written telework agreement to ensure that telework does not diminish performance.

- Provide an interactive telework training program for employees eligible to participate in telework and their managers; this training is to be completed prior to the signing of the telework agreement.

Telework is a requirement of all Agencies' COOP priorities.

Additionally, Agencies were required to set telework participation goals and assess the impact of telework on areas related to Agency efficiency, effectiveness and sustainability.

GSA's Plan

In December 2010, GSA Administrator Martha Johnson appointed Sharon Wall, Federal Acquisition Service (FAS) Assistant Commissioner as the Senior Telework Accountability Official to lead GSA's effort to meet the new goals outlined in the Telework Enhancement Act of 2010. Under the supervision of Martha Johnson, and oversight of Sharon Wall, the Telework Program Management Office (PMO) was launched in January 2011.

The PMO was comprised of GSA employees from the Public Building Service (PBS) and Federal Acquisition Service (FAS) who had a keen interest in Telework initiatives, but were also subject matter experts (SMEs) in different areas that Telework encompassed. Many of the participants were graduates of GSA's Advanced Leadership Development Program (ALDP), while others were Interns or other employees whose skill sets matched those needed in the PMO.

Six tranches were formed to address the Telework initiatives:

1. *Performance Measurement*
2. *Policy*
3. *Skills Development*
4. *Technology*
5. *Customer Service*
6. *Communications and Marketing*

Each tranche worked to guide the expectations and responsibilities of managers and teleworkers, enhance/support the program, and establish awareness and appreciation of the benefits of telework.

In determining a path for the Telework PMO to follow for this aggressive initiative, the team defined a strategic game board, which showed the current state, case for change, and future state, connected with strategies and barriers. With this game board in place, the PMO began its transformational work.

The Business Case for Telework

Besides the mandates of the Telework Enhancement Act of 2010, telework makes sense for GSA, and other Agencies in many ways as it helps in the achievement of other key goals. The business case for telework includes the following:

Creation of efficiencies and cost savings = savings to the taxpayer

- Employees save on commuting costs and Agencies save on Smart Benefits costs. This DC Metropolitan area program allows Federal employees in the DC area to receive a stipend for approved commuting costs on Metro bus, Metro rail, Commuter buses, and other approved local rail service.

- More employees working from an alternate location means less need for real estate, and lower operating costs for Federal buildings.

- An experienced mobile workforce can continue operations in an emergency, thereby not losing productivity during emergencies where the Federal Government is shut down, or conditions make it difficult to get to an office.

Protecting the environment

- Leads GSA closer to a Zero Environmental Footprint (ZEF) state.

- Fewer cars on the road mean less gas consumed, and less pollution.

- Decreases Greenhouse Gas (GHG) Emissions for Scopes 1 and 2.

Accommodating the changing needs of the workforce

- Workplace flexibilities have changed; the corner office is no longer the norm. Employees can be engaged and collaborative in an open and changing environment, which is supported by telework.

- Flexibility attracts forward thinkers to government.

- Flexibility reinforces team collaboration and effectiveness.

Creation of an innovative government of the future

- Mobile work supports real time innovative customer solutions.

- IT has become the backbone of mobile work- as employees need to work in alternate locations, IT makes them more effective.

Improving employee performance

- Employee performance is enhanced by being in a conducive environment and having options for collaboration and different work styles.

- Today's work is different; there is less need to be "watched", and more need to be connected.

- Telework sets the stage for managers and employees to communicate more.

On a larger scale, telework complements other initiatives put forth by the Federal government. Some of these initiatives include:

- **Executive Order 13514:** Federal Leadership in Environmental, Energy and Economic Performance

- **Executive Order 13078:** Increasing Federal Employment of Adults with Disabilities

- **Presidential Memorandum** (June 10, 2010)**:** Disposing of Unneeded Real Estate

The Telework Enhancement Act of 2010, Executive Orders, and the business case all support an alternate work situation. Breaking down barriers and dealing with change management is not mandated, but in order to be successful, management supp buy-in, and encouragement are key.

Preface

The PMO realized the challenge of implementing and enhancing telework, and created this Recipe Book to share GSA best practices from the work of the six tranches. This is certainly a work in progress, and even with the best of intentions, some initiatives do not go according to plan. In the six month period of time (January to July 2011), when the PMO laid the foundation for its most transformational work, there were many bumps in the road. The telework transformational journey is retold in this book with information gained through discussions, notes, interviews and feedback from the tranche members. Each section, or recipe is told in a slightly different tone, which represents the way the tranches' "personality"This recipe book is useful for anyone who is dealing with the same challenges in implementing or enhancing Telework initiatives. Not all of these approaches will work for all Agencies, but some strategies and tips may mix well with ideas already in place. Each "recipe" is told with the following "ingredients": *Vision*, *Challenge*, *Process*, *Players*, *Outcome*, and *Next Steps*. Additionally, answers to frequently asked questions (FAQs) are provided, along with a guide to help you choose Subject Matter Experts (SMEs) to work on your team.

Performance Measurement

The Vision

The Performance Measurement Tranche's vision was threefold:

- Align telework with mission-centric business intelligence that informs decision-making.
- Transition measurement of telework from static paper telework agreements to a system of automatic measurement- The Telework iDashboard.
- Create Agency-wide transparency of telework data.

The Challenge

The team challenges were deeply rooted in telework data that had no baselines or accuracy. Telework hours were reported on paper telework agreements, and in some cases in the Electronic Time and Attendance Management System (ETAMS), however these numbers were seldom verified. Many of the telework agreements were out of date and did not reflect what the employees were doing; in many cases, the agreements showed that employees had specific telework days, but they weren't allowed to telework. Additionally, there was confusion about what type of telework was being done, and there were 4 telework codes in ETAMS.

These issues led to the greater challenge of developing a fully automated tool to track the telework data, and then using that data to measure (1) the correlation between telework and productivity and (2) telework and customer satisfaction, since no measures had been previously established.

The Process

To meet the statutory requirements stated in the Telework Enhancement Act of 2010, the team identified the following areas of proposed measurement and case studies to focus on:

- **Telework Achievement Measurement:** Measure the actual hours of telework per full time employee (FTE) per pay period.

- **Employee Engagement and Satisfaction Measurement:** Add statements to the OPM Employee Viewpoint Survey and GSA's Q-12 Survey.

- **Sustainability Measurement:** Measure various benefits to sustainability from telework to include (1) avoided employee commuting emissions, (2) employee commuting costs savings, and (3) travel data.

- **Technology Performance Measurement:** Measure (1) the ease and satisfaction of technology while teleworking thru a question added to the OCIO Customer Satisfaction Survey, and (2) reduction in Help Desk volume inquiries pertaining to teleworkers.

- **External Customer Satisfaction Case Study:** Customer service needs to be seamless regardless of where the GSA employee is working; case study on the correlation between telework and customer satisfaction.

- **Workforce Productivity:** Employees who work remotely should not be measured any differently than those that are in the office; White Paper developed to discuss this.

- **Space Utilization:** GSA's Extreme Challenge- Central Office Building at 1,800 F Street. Once renovated, the building, which used to house 2,400 employees, will have over 4,000 GSA employees from the National Capital Region. Real estate savings will be tracked, and can be used with telework metrics.

ETAMS was the obvious choice as the tool to accurately measure Telework Achievement, as opposed to obtaining the data from static telework agreements. Telework codes were already part of ETAMS, and in some regions of GSA, the coding was used accurately to capture actual telework hours, so this was an obvious solution.

To measure Employee Engagement, four statements were added to the Office of Personnel Management's (OPM) Employee ViewPoint Survey:

- Leaders encourage and support employees that telework.

- My organization provides the appropriate technology necessary to telework.

- I am able to receive the same growth opportunities in my work group when I telework.

- Teleworking opportunities provide a better work/life balance.

For GSA's Internal Q-12 Survey, the statement used, "Leaders encourage and support employees that telework," was from the bulleted list on the previous page.

In the area of Sustainability, a major focus was on measuring the avoided employee commuting emissions. A question was added to the OCIO Customer Satisfaction Survey to measure the ease and satisfaction of technology while teleworking.

Finally, case studies were developed for Customer Satisfaction, Workforce Productivity, and Space Utilization. By studying all of these measurement areas, the full benefits of telework may be captured and realized by the agency.

The Players

The success of the Performance Management Tranche would not have been possible without much collaboration with a variety of offices. Below is a list of those who provided valuable insight:

- **Office of the Chief Information Officer (OCIO):** Thru them, the telework survey question was added to the OCIO survey for GSA employees.

- **Office of Budget and The Chief Financial Officer:** Assisted with the analysis of survey data, and provided expertise on performance measurement.

- **Office of the Chief People Officer (OCPO):** Instrumental in getting the questions on the OPM Employment Viewpoint and Q-12 Surveys, which measured employee engagement.

- **Office of Real Property Asset Management:** Provided expertise in the area of space utilization measurement.

- **GSA's Administrator's Office:** Provided a "sound board" to ensure that the efforts of the Tranche were not being duplicated elsewhere in GSA.

- **The Gallup Organization:** The organization that provides survey support to GSA; they were consulted to help with the wording of the Q12 Telework question. www.gallup.com

- **Regional Offices:** Employees from all over GSA participated in the Customer Satisfaction Case Studies, and in the Productivity Case Study Trial.

- Other Agencies were contacted to provide examples of best practices in measuring telework.

The Outcome

The Telework iDashboard was developed and launched. The team leveraged a product that was already in the Office of the Chief Information Officer (OCIO), and the licenses were already purchased. This tool pulls telework data from ETAMS and provides updates on telework hours worked, with added metrics for travel cost savings. The iDashboard also breaks the data down by region.

The use and importance of two telework codes to track time in ETAMS was established and promoted to all employees to ensure accuracy and transparency. The codes in use are Telework and Emergency Telework. There was much analysis and research into the other areas of measurement as well. Many of the results are forthcoming.

The Next Steps

- The Dashboard 2.0 (phase 2 of the iDashboard) is currently in the works and will be launched early in FY12. The tool will be transitioned over to the OCFO Budget Team who will continue to collect the data. The data will be housed in Access rather than Excel to manage the large data set. The layout of the dashboard will change to incorporate Travel data as well. Through the use of the "Emergency Telework" code, big events such a black out will be easily recorded. Dashboard 2.0 will be connected to the other enterprise wide measures such as printers, travel reduction, and strategic sourcing. The new dashboard will contain historical figures, by region and fiscal year, a change from the original which separated historical and current data, and were recorded by calendar year.

- Results of the OPM Employee ViewPoint and Q-12 surveys will be provided in September 2011. Gallup will provide a driver analysis on the Q-12 survey for each workgroup. This will enable the agency to measure employee engagement as it relates to telework.

- The Office of Chief Information Officer (OCIO) survey will be launched in October 2011, with final results to be provided in January 2012. This survey will demonstrate how satisfied employees are with technology while teleworking.

- A Productivity case study will be done in GSA Regions 5, 6, 7 and The Administrator's Office (DC), with results expected in November 2011.

- Customer Satisfaction case studies are being done, utilizing qualitative and quantitative analyses.

- GSA's internal The Mobile Work forum was used to identify GSA groups that were enabled by telework to better serve their customers. Five groups were identified and interviewed. The case study focuses on the results of the interviews and also provides analysis, best practices, and recommendations.

- Quantitative research is currently being conducted to analyze the relationship between the historical **PBS** and **FAS** customer satisfaction surveys and mobile work. Results will be provided by the end of August 2011.

Policy

The Vision

GSA had a robust telework policy in effect since 2008, but with the requirements of the Telework Enhancement Act of 2010, GSA needed to "reboot" its policy. Per the requirements, telework was to be a part of all Agency COOP priorities, and goals had to be set for telework participation and its impact on areas related to agency efficiency, effectiveness and sustainability.

The Challenge

One of the biggest challenges with the new policy was to make it aggressive enough to thrust GSA forward as a leader in Telework and also make employees feel empowered to Telework, yet controllable enough for supervisors to feel their rights and power have not been diminished. In finding this right mix, inevitably, the shift would have to ensure that every-one realize that Telework and performance are not the same, and should not be addressed as such in the Telework Policy.

The Process

Drafting the updated Telework policy was an evolving process involving input from all levels and disciplines within GSA. The policy is a living document and is to be a catalyst for making GSA a better place to work and better able to achieve its mission and customer needs. GSA is a customer driven agency and an employee friendly agency. Keeping this in mind, the following tasks and accomplishments show the journey toward the updated policy.

- Selected highly motivated key individuals with progressive thinking to develop bold policy changes.

- Team members participated in two "Leadership in a Minute" videos focusing on workplace diversity and teamwork. These videos were geared towards all employees.

- Highlighted policy changes at Central Station Meetings. These meetings were open to all GSA employees who had an interest in participating and/or providing feedback to the policy process.

- Requested comments and input from GSA senior management and one GSA Region (R6) Telework supporters during the early stages of policy development.

- Developed a Workplace Flexibility document to be used for the White House Working Group, which supports telework and other initiatives that support women and families.

- Participated in pre-decisional meetings with Union Representatives from AFGE and NFFE regarding bold ideas for the policy.

- Sent the draft policy through the Red Border process for GSA-wide comments.

- Participated at the GSA Expo Telework Booth to promote Telework government-wide.

- Team member presented information on GSA's proposed policy at a nationwide Federal Executive Board meeting.

- Reviewed and provided input on the GSA mandatory Telework training script.

- In the process of preparing a Desk Guide to Telework to ensure consistent implementation of the policy GSA-wide.

The Players

Creating a bold, rebooted version of the Telework Policy would not be possible without being able to collaborate with different groups within GSA. As Telework touches every facet of GSA, the players had to represent different facets of GSA. These included Human Resources; The Office of the Chief People Officer, The Office of Government Wide Policy, Agency Clearance Officers, and the Office of the Chief Information Officer. The Union (NFEE and AFGE) Representatives were contacted for pre-decisional comments, however, consideration could be given to inclusion of these representatives on the PMO from the beginning.

The Outcome

The policy that was put forth had some bold changes that included:

- Allowing all employees to telework with limited exceptions.

- Utilizing only 2 codes for telework- telework and emergency telework.

- Removal of telework schedule requirements.

- Allowing Union Representatives to telework.

- Adding requirements for all meetings to have virtual attendance capabilities.

The Next Steps

The Policy Tranche will continue working with GSA senior management and the Office of General Counsel for final concurrence on the Telework Policy. A final negotiation with AFGE and NFFE is the last step before the new launch. The Policy will be rolled out nationally, with a mandatory training and a Desk Guide for implementation.

Skills Development

The Vision

The Skills Development tranche was tasked to expand individual and organizational capabilities, and to deliver a path forward to assemble the requisite tools, resources, technology, and training to set the standard for effective workforce mobility and cultural transformation. The tranche worked to identify the skills enhancements, experiential learning opportunities, and knowledge management strategies necessary to eliminate cultural barriers to mobile work – and sought to connect the dots between existing information, ideas that emerged during the life of the PMO, and new training to be put in place. Ultimately the tranche deliverables should lead to enterprise-wide awareness, and use of, resources, learning environments, and knowledge transfer activities to close the skill gaps and achieve a transparent, innovative mobile work environment across GSA.

The Challenge

The challenge was to support achievement of an enterprise-wide culture transformation and convey the message that telework is a team sport. Our goal is to ensure GSA has the right tools and resources to enable its workforce to innovate and embrace experiential learning while moving along the path toward organizational excellence. We sought to support the efforts of the Telework PMO to help identify workplace innovations to empower individuals and teams to work collaboratively and influence GSA's strategic direction. Of particular concern was making sure the group kept pace with constantly emerging ideas and approaches to improving mobility.

The Process

The Skills Tranche embarked on a path to make it easy and efficient to "just work", and to expand organizational and individual capabilities. Our goal was to help evolve our work practices to a point where teams depend on their members no matter where they are, and that they can work effectively by reaching out, networking, maintaining relationships, and staying connected from anywhere. Simply, our approach was not to reinvent the wheel, but to build upon the wealth of great ideas, make them accessible, and create supplemental resources as necessary.

The team evaluated existing courses offered through GSA On Line University in support of telework activities and determined that we needed to develop a customized mandatory training course that would advance the Administrator's organizational transformation goal. The course will help the entire GSA organization understand the importance of virtual, mobile, flexible, and telework. The course was developed in house – written, produced, directed, acted, filmed, and edited by GSA! To complement the mandatory training that touches on the philosophy and importance of telework, the team is in the process of procuring training to teach individuals how to manage effectively in a mobile work environment, and to provide downloadable learning aids and how-to guides for effectively managing teams in a virtual environment.

The team worked to build upon existing resources, tools, training, and experiences, and to leverage the use of communication tools, such as the Telework Forum/Mobility page on GSA Insite, as a repository or gateway to readily access information. The Telework Forum/Mobility page was developed by the GSA Telework PMO as the go-to web site for information on the workings of the Telework PMO, and houses a wealth of tools, resources, videos, and references.

The Skills team used a GPM (global project management) approach and drew upon the team's knowledge of systems for data collection and sharing. The approach was collaborative and inclusive. We used established relationships and built new relationships to leverage individual team member strengths and experience to complete this project.

The Players

The Skills Tranche was fortunate to have input and participation from across the entire GSA organization. The team drew upon resources from the Federal Acquisition Service (FAS), the Public Buildings Service (PBS), the Office of the Chief Information Officer, the Chief People Officer (CPO), the Administrator's Office and related staff offices, the team also work closely with the folks from GSA On Line University (OLU), an in-house education vehicle that houses web-based, electronic training for GSA personnel. The team infused a measured blend of the unique training expertise in the FAS, PBS, and CPO organizations into its work. The team comprised members from across the entire nation and multiple regional offices.

The Outcome

GSA Online University (GSA OLU) will offer mandatory telework training and managers training. There is a plan to upgrade the existing Telework 101 courses for managers and employees based on the new GSA Telework policy. The current Telework Forum/Mobility page on GSA's Intranet will include links to other mobile/ virtual work resources and guidance, such as how-to videos, recommended reading, how-to guides, facilitator guides, mobile work tools, and other useful information.

The Next Steps

The team will continue to finalize its products, work with the Communications team to incorporate the work of the Skills team into the Telework Forum/Mobility page, and develop recommendations for future improvements that can be implemented by existing GSA organizations.

Customer Service

The Vision

Methods of supporting the customer should not change because an employee is teleworking; GSA customers should receive the perfect customer experience regardless of where the employee is working. The Customer Experience Tranche envisioned all GSA employees understanding how to create the perfect customer experience while teleworking by using all the available tools, resources and technology to deliver seamless services.

The Challenge

The two largest GSA service organizations; Federal Acquisition Services (FAS) and Public Building Services (PBS) have two very different missions and customer bases. The team's challenge was to look across all of GSA and understand how it interacts with customers and leverage the many great things that were already happening, as well as capture best practices from outside the organization and convey them to all GSA employees.

The Process

- **Data Collection**
- **Information Sharing**
- **Support and Training**
- **Measurement**

The team began by defining its focus on the external customer. GSA's external customers are federal agencies and vendors that rely on GSA for service and/or business. Also recognized was that communication and collaboration with numerous internal customers has a domino effect on the external customer as well. Therefore, the good practices that were used to support the external customer could be used to support the internal customer as well.

The first step was data collection; reaching out to the organizations and capturing all of the best practices employees and managers were currently doing to support customers. The team's methods for data collection are listed below:

- Conducted two focus groups of experienced teleworkers; One for managers and one for employees.

  •• Utilized technology to bring representatives from across the US together for the virtual focus groups.

- Captured employee's telework experience during Super Bowl week in Fort Worth, TX, and during Telework week. This was done because all employees in the Fort Worth area were affected by the traffic and street closings during Super Bowl week, so they were highly encouraged to telework.

  •• Established an email account for employees to share their stories.

- Conducted internal agency surveys capturing current telework experiences.

- Collected personal stories to be posted on an internal Telework Forum page. http://insite.gsablogs.gsa.gov/telework

- Researched telework programs and success stories from other Federal agencies and private businesses.

The team took the best customer support methods collected and created a *Telework Tool Kit*. The tool kit was designed to provide basic information to teleworkers to help them prepare to telework and specify methods to use to stay in communication with the customer and their team.

The Customer Experience Tranche was spotlighted on the Telework Forum for two weeks. The team took this opportunity to emphasize the many ways to support the customer. A video of an employee mobileworking was the lead story on the forum page. This showcased another method of providing direct support to the customer while mobileworking. Experiences collected were crafted into stories and shared on the forum page. This provided the agency an opportunity to read about good practices, comment, and share their own stories. Not all comments were positive. Employees had legitimate concerns with staying available and connected with customers while teleworkers. Members of the team commented by sharing instructions and guidance on the use of technology and improvements planned for the near future.

The team also looked for ways to provide direct customer service training to employees. It linked up with the Skills Tranche and provided an outline for customer service training to be added to the employee and manager training; this is an ongoing process.

The team brainstormed other ways to keep the customer on the minds of teleworkers. In support of rolling out the employee training, a small media campaign was developed. Two posters showing various ways to collaborate and support the customer were developed. A third one is in design. The posters were turned over to the Marketing and Communication Tranche to be used in support of future messaging.

The team participated in the 2011 GSA Expo by joint hosting two focus group sessions with federal agency employees with the Telework Exchange. These focus group sessions were an excellent forum to share the GSA story while helping agencies explore ways to move their telework program forward

The final area of focus was to measure customer satisfaction. The team joined forces with the measurement tranche to find links between increased teleworking and customer satisfaction. The team is using qualitative and quantitative methods to explore this link. Work on the customer satisfaction measure is ongoing.

The Players

All employees and customers were essential to the success of this team. The focus groups with managers, both novice and experienced teleworkers provided valuable insights, and helped to develop the Telework Toolkit. Employee feedback was also key in understanding where GSA really was on the acceptance of telework.

The Outcome

There have been many great results from this process:

- Dialogue is continuing on how to support the customer while teleworking.

- Employees are more engaged in the telework process.

- Employees are more cognizant about measuring telework and customer service.

- Tools to support telework are readily available.

- Employees continue to share stories, challenges, and best practices on the forum page.

The Next Steps

The final media posters were completed and turned over to Marketing and Communications to support future messaging. The team will also work with the Skills Tranche to support their efforts with employee training. Servicing and supporting customer intimacy must be reinforced as a priority for all teleworkers, so the team will be working with Marketing and Communication on a plan to ensure that customer support dialogue continues GSA wide. Work will continue on the Customer Satisfaction Measure and until all deliverables are met.

Technology

The Vision

The Technology tranche of the Telework PMO realized that it needed to work with the Office of the Chief Information Officer in order to drive change into the organization. Because Information Technology is critical to employees' being able to work remotely, GSA wanted to ensure that OCIO's efforts were in line with the transformation that telework represented.

The Challenge

GSA employees already had a range of modern technology available to them, such as **VPN** software, laptops, and "softphones" (i.e., telephones able to be used on a laptop). However, employees are spread across a spectrum of technological skill. Some employees are digital natives and can easily figure out how to use these tools; others need more assistance, or may resist using them. The challenge, therefore, was making sure that the IT infrastructure could handle increased teleworking, and that employees had the right tools to do so.

The Process

A key event in the Telework Tranche's efforts was the preparation for Telework Week. To prepare for the increase in the number of teleworking employees, the Technology tranche engaged the IT Help Desk staff to collect data from the employees who called in with telework IT questions. This data was useful in two ways:

- It allowed us to identify areas for improvement in network infrastructure.

- It informed us as to what employees were having trouble with as they used the teleworking technology.

21

Analyzing this information on a nightly basis during Telework Week helped the Telework Tranche adjust its strategy to making telework more widespread across GSA.

The Tranche incorporated this information into a set of recommendations to the Office of the Chief Information Officer. These recommendations encompassed provisioning, training, and other upgrades.

The Players

The Technology Tranche collaborated with people with and without an IT background who had an interest in GSA IT Solutions. Additionally, members of The OCIO infrastructure operations management team partnered with the team to implement new and enhanced technology solutions.

The Outcome

The Office of the Chief Information Officer has implemented our recommendations to varying extents. In the transition to the Google mail and collaboration platform, the benefits of that platform to telework were emphasized to GSA employees. The Google platform is perfectly suited for telework and collaboration, with its features like presence status and video chat. Telework has been fully integrated into GSA's technology strategy.

A bold statement requested by the Technology Tranche and adopted by the OCIO was an organizational change to integrate the VPN and Citrix teams under a single management umbrella called the Workforce Mobility Staff. The new team is solely dedicated to enhancing and implementing virtual employee technologies and making the virtual/mobile employee's experience as productive as it can be.

The Next Steps

Now that Telework is part of how GSA carries out its mission, GSA can lead other agencies to that end. Other agencies have approached GSA, interested in the best way to increase the use of teleworking among their employees and the best IT solutions to use. Using the knowledge gained through our experiences, the Telework PMO will spread telework government-wide.

Communications & Marketing

The Vision

To develop a comprehensive marketing and communications campaign built upon a consistent enterprise brand, which would be self-sustaining after the dissolution of the PMO. In a successful end-state, every GSA employee will know the business case for telework.

The GSA Telework Program Management Office (PMO) is responsible for outlining the scope of the movement toward telework and remote collaboration at the General Services Administration (GSA). It was the Communication and Marketing team's responsibility to develop and spread the case for these movements through strategic messaging, collaborating with the other PMO groups, and leveraging private industry.

The Challenge

The biggest challenge for the Communications group was shifting from the prevailing ideas of telework to the consensus understanding that telework is a tool that allows GSA the flexibility to excel in a mobile work environment, become a government modeling environmental stewardship, and linking to employee performance.

Telework was somewhat negatively viewed by many managers as:

- Providing an easy work day with potential drops in productivity.

- Working at home (rather than in locations outside the office which may or may not be at home.

- A lessening of management control characterized by questions like, "How will I know what someone is doing if they aren't in the office?"

Telework was less negatively viewed by employees who were of the opinion that it provided better work/life balance and a potential increase in productivity.

The issue was finding the path to reconcile the perceptions in a way that advanced telework. It was achieved through effective communication strategies of the business case that resonated with the diverse groups within GSA.

Part of the communication strategy was to encourage Agency-wide collaboration as a platform for change in a culture of silos and historical independence of work areas.

The Process

In alignment with the agency wide initiative of transformation, the Marketing & Communications team identified audiences, developed and refined messaging, and utilized both existing and new channels that encouraged a high level of collaboration among GSA employees. The communications plan was developed to address the needs of several different internal and external stakeholders and provide for the dissemination of accurate, timely and consistent information.

Internally, we focused the efforts for information and collaboration into one centralized platform to push people towards and disperse information from the Telework Forum. This forum reached all of GSA and was the platform where the PMO shared success stories, housed measurement data (Dashboard), PMO accomplishments, highlighted pilot programs, how-to training videos, and collaborative discussions about successes, barriers, concerns, and potential solutions.

Various channels of communication were used:

- "Telework Is..." e-mail series from Sharon Wall Telework PMO Mailbox.

- Telework Forum on GSA's Intranet.

- Messaging Map for executives and telework advocates.

- Simple key messaging like "Work is what we do, not where we are!"

- Telework booth at GSA EXPO in California in May 2011.

- Telework PMO-hosted, Central Station calls to build a cohort of telework advocates, (not necessarily part of the Telework PMO) supporting each other, and providing feedback.

- Partnering with Telework Exchange (www.teleworkexchangeorg) for Telework Week.

- Focus Groups for both internal and external stakeholders.

- Presentations for external stakeholders.

- Speaking engagements for external stakeholders - such as FedNews Radio, The White House, and other Agencies.

The Players

The Communications and Marketing team relied on the expertise and collaboration with many groups. GSA's Office of Communications and Marketing was integral in developing concise, clear and unified messaging. They worked hand in hand with the GSA Public Affairs Officers in all 11 regions to ensure that all employees got the same information in a timely manner. During Telework Week (Feb 14-18), The Telework Exchange was a key driver in helping to create metrics for telework participation across GSA. The Federal Acquisition Service (FAS) Office of Administration and Office of Strategy Management also provided input and collaboration during Telework Week.

In order to create the best marketing and communications strategies for an initiative like telework, it's important to include Agency-wide communications team to ensure that the messaging is relevant, cohesive and timely, so that it well received by all employees. Communications like this will have a bigger impact if it is perceived in a positive light and without confusion.

The Outcome

The Communications and Marketing team was able to deliver clear, consistent and effective communication/engagement vehicles capable of raising awareness and educating all levels of GSA employees and external stakeholders about agency wide telework initiatives that will enable those employees to work more productively, live more sustainably, and protect more vigorously the integrity of the federal government.

The consistent and constant messaging across various platforms transformed the way stakeholders understand the telework business case from the baseline listed in challenges to our desired goals and outcomes.

The Next Steps

There is still work that needs to be done, especially externally. Telework is not something specific to GSA and speaking engagements and interest from external agencies are mounting. With the messaging created, white papers written, presentations created, and previous speaking engagements already completed, it is leveraging all the information stored in a central location to meet the future needs of external communication.

Where do we go from here?

From the inception of the Telework PMO in late December 2010 to now, a lot has changed, and a lot has been accomplished. GSA was able to use its most valuable asset, its employee base to make transformational changes by thinking outside the box and viewing telework, not as a stand-alone initiative, but a driver to so many other sustainable requirements. The GSA Telework PMO is not done yet- the members may change to allow for an influx of new ideas, but the creativity and drive will remain. When transformational change occurs, it has to continually evolve to be effective, and that is what will continue at GSA.

Training will be rolled out in conjunction with the bold Telework Policy, IT will continue to evolve (GSA has moved to the Google cloud where collaboration is second nature), telework goals will be measured and targets will change, communications will continue to highlight the message in different ways, and enhancements will make business more efficient.

The creation of a Telework PMO will benefit your Agency in many ways, and will be the impetus for change. The story told in this rec- pe book can help get you started in the right direction, or you may find some ideas to tweak what you already have.

Further information on GSA's Telework initiatives, as well as training, articles and contact information may be found at: www.gsa.gov/telework.

Frequently Asked Questions (FAQs)

The GSA Telework PMO had many interactions with other Federal Agencies and Private Organizations. Below is a list of the FAQ's by category:

Project Organization

What are your deliverables?
How did you determine these?

By the end of the PMO, the team hopes to have a (1) robust toolbox of videos, stories, and lessons learned that can be used to help employees and managers adapt to mobilework, (2) a new Telework Policy in place, (3) Training Modules, (4) Data to show measureable success and baselines to measure future success. These deliverables were determined mainly by the Telework Enhancement Act, but also by the needs of GSA at the time. These deliverables are just a starting point, and will change over time.

How do you begin the cultural shift from traditional ways of office based working to teleworking?

The cultural shift may be gradual. Keeping a constant focus on nudging towards the goal is essential and it must be constant. The notion of telework being a team sport is key; if one person teleworks, then the rest of the team can be considered teleworkers as well (at a minimum because they are adjusting their work style to work with the teleworker). Getting support from upper management and having leaders demonstrate telework practices will shift things in the right direction.

How do you handle the ups and downs of design and delivery?

This is handled in more or less the same way that normal work is managed by managers skilled at managing a remote workforce. There are regular group check ins with team leads (weekly). Team leads then update their team members. Project measures are tracked and reported, and any deviation from the schedule is addressed in project update meetings. Community is built by on-going informal connections through email, instant messaging, phone contact, and so on. Handling the ups and downs relies on having strong program governance and an openness to enabling team member autonomy and collaboration.

What is role of the Subject Matter Expert (SME)? Do they make the final decisions?

See the Appendix for a Recipe for Subject Matter Experts. Generally speaking their role is advisory.

Policy

How do you make a telework policy flexible enough that it does not become ambiguous?

By using the flexibility given in the GSA policy, business lines can create office situations that match the customer needs, and the team. Policies need to be for the group, not mandated on individual concerns, or individual performance concerns. Performance management should not be incorporated into telework; that belongs at the performance management levels/systems.

How do you deal with those positions that are not applicable for Telework in the same way as others?

Essentially, all of GSA's employees are eligible for telework; however there are some positions that require a presence in an office or building. Even with that being the case, there are opportunities to telework at locations other than one's home. For example, a Property Manager can telework in another building while visiting a tenant, or a Project Manager can telework at a construction site's office. IT tools make it easier to stay connected to teams and customers, so almost all employees have opportunities to use them.

Information Technology

How do you separate telework tools and technology from performance management perceptions?

Telework tools and IT are to allow for increased productivity, mobility, flexibility and collaboration, which would ultimately lead to increased performance, but that management is clearly defined by other tools. Clear performance metrics based on results have to be in place from the beginning.

How do you make IT support easy and cost effective, especially for enhancements (i.e. file encryption, video conferencing, etc)?

IT has to become a partner in telework by supporting a robust platform that will provide consistent and reliable tools. The creation of self help tools and the roll out of effective training will help employees rely on themselves. Other ideas include IT Open Houses, Hands on Training roll outs, and collaboration within business lines to share equipment and IT solutions.

Skills Development

How do you roll out training uniformly with regard to different learning types? Training is mandated in the policy, but how do you make sure all employees "get it?"

Employees are different and they learn differently as well. Training has to be diverse as they are so no one feels intimidated by the process. Offer many methods, such as Do-It-Yourself IT (as noted in GSA's Telework Forum), Train the Trainer programs, IT Open Houses, Online FAQ's, etc. Ensure the message is consistent and roll out training gradually.

Customer Service

How do you create a seamless customer experience across an Agency when you have different customer bases?

One approach will not fit all. Customers require different levels of intimacy. This should not be seen as a barrier to telework, rather an opportunity to engage the customer in the telework discussion to find mutually agreeable solutions for both.

Communication

How do you maintain strong team relationships when teams Telework?

There are many ways to keep a team connected, even with collaboration tools like Sametime meetings, Sametime IM, and videoconferencing. GSA has partnered with George Mason University on a study about work attitudes in a telework vs., traditional office environment. The study will focus on 2 kinds of social connectedness; behavioral (face to face interactions) and perceived (how the employee feels regardless of face to face interaction). The study outcomes will be used to develop employee engagement and retention practices and programs.

It seems that your Administrator, Martha Johnson was on board, and a strong supporter of telework. How do you get Stakeholder buy-in?

As with any project there must be a 'burning platform' that others understand and support. In this case it was the Telework Enhancement Act combined with requirements to save on corporate real estate and to reduce carbon footprint. Keeping key stakeholders informed is critical, and giving them a specific role also helps. For example Martha Johnson was invited to blog and present short videos on various aspects of teleworking.

How do you maintain connections to other organizational projects and initiatives?

This requires people on the ground making connections, being alert to overlaps and duplicated effort, having the confidence to question things that seem disconnected or out of touch. The way this project was designed, drawing on expertise from across the organization, facilitated our ability to make and maintain connections with other projects and initiatives.

How do you disseminate information to all employees and ensure that the messaging is cohesive?

Messaging is handled through a single point of contact that checks for consistency of language and content. In a project of this scale it is critical that there is no dilution of the message. Having a specifically designated internal website was instrumental in gathering and disseminating information in a consistent way. Employees were invited to participate in the website discussions and to look there for answers to their questions.

Management

How do you deal with "trust" issues? How do you create a more effective paradigm shift for first-line supervisors who may have control tendencies, thereby making them reluctant about endorsing the telework initiative to its fullest effect?

1. Consider telework as one tool for overall succession planning -- a manager should feel comfortable enough to telework knowing that his/her teleworking allows someone else the opportunity to assume more of a leadership/authoritative role at the office in his/her absence. From a subordinate perspective, might allow a manager not teleworking more 1:1 time with those employees who are not teleworking, which helps with professional development as whole.

2. Focus more on the work output -- if performance measurements are being achieved and work expectations are being met, why does there need to be so much emphasis on time and attendance at the workplace? Those managers unwilling to promote teleworking are likely struggling with control issues. There are, of course, specialized circumstances that might delay more regular telework -- e.g. newer employees with certain job functions that require more training in a group setting -- but professional development at the early stages depends even more on the coach/employee.

3. Training for Managers - the educational process is critical for managers. This training cannot solely consist of independent on-line training, since the collaborative dialogue comprised of shared telework feedback, which GSA strongly encourages, occurs best in a more traditional classroom format.

Based on experiences so far, what are some tips on managing telework?

- Collectively address questions about telework - lose the ambiguity, be transparent.

- Manage telework as a business practice, as opposed to an individual initiative.

- Develop telework policies that work w/in your business line, but keep the general GSA Telework Policy as the main model.

- Let the IT Help Desk be your partner; ensure IT is consistent.

- Monitor performance across the entire business line.

- Build in community events.

- Give regular, timely performance evaluations to individuals.

- Appoint someone to oversee telework hoteling and alternate workspaces.

- Maintain enthusiasm, commitment and engagement.

Appendix

The Vision: SME role

What is a 'Subject Matter Expert' (SME)?

In a project to develop telework skills, knowledge, and behavior, it's easy to assume a nebulous vision of a telework guru going around giving ad hoc but sage advice to hard-working project team members and then seeing them act on it.

But a workable vision for SME value-add to a project is much harder edged than this. Envision an effective SME. He/she has in-depth, specialist or expert knowledge of a business area, work process or system functionality. With this goes the ability to transmit and share his/her knowledge to a project team in a way that helps them successfully meet, or even exceed, their goals and objectives.

For example, a measurement SME will be able to help the Measurement Team choose specifically, what to measure, why to measure it, and how to measure it.

The Challenge:

There are several challenges to the SME role:

- The SME brief is not clear so he/she doesn't know what the expectations are in terms of contribution and delivery.

- The project team does not recognize the need for SME support in the tranches (or for a cross-cutting SME for example for change management).

- The program lead does not have the skills or resources to select SMEs.

- The team members do not know how or when to ask for SME support and assistance.

- There is an inadequate match between what the team needs and what the team wants from the SME – are they looking for a trainer, peer reviewer, approver, knowledge sharer or something else.

- There is no point of contact for the SME to report or refer to for guidance and updates.

- SMEs are not perceived as a 'real' contributor and are left off of communications and out of meetings that could be relevant.

- The SME has other organizational roles that take precedence over this one.

The Process

Making the SME role successful for the project requires, as a first step the development of a clear brief for the role. This should include information under the following headings:

Heading	Example text
What we are looking for in an SME	We are seeking someone who can bring technical expertise to our Policy Tranche. Ideally you will have worked with the Agency's telework policy and be well versed in the Government's stance and direction on telework policy. You will have built a reputation as a 'go-to' person in this expertise, and be able to give advice and direction on how we can extend and develop telework within the bounds of the existing policy. We also expect you to have the skills and confidence to both recommend and push for policy changes if it becomes evident that these are needed. Additionally we are looking for someone who is comfortable with sometimes chaotic, emerging situations, who can be proactive and is quick at 'connecting the dots'.
What you can expect from us	A challenging project that provides growth and learning for you as we take teleworking to new heights. Great people to work with.
What we expect from you	25% of your time for the duration of the project. Proactive involvement, suggestions, advice, recommendations as the project progresses. New ways of thinking and new perspectives on your domain of expertise as it applies to teleworking. Best practices from other organizations. Full participation as a team member. Contributions to our knowledge bank (white papers, articles, other resources)

Activities	*Participation in weekly team meetings*
	Regular contributions to the project collaboration space
	Development of team member skills and expertise (e.g. through running lunch n learn sessions)
Deliverables	*To be agreed*

The second step is selection of the right person or people. In a telework project several SMEs are required. The idea is to have one attached to each tranche plus one for communications/stakeholder engagement, and one for change management. An article by Jose Fajardo lists some SME selection criteria (adapted below):

Selection Criteria	Where to Probe
Domain expertise	*Does the person have deep expertise in the specific domain e.g. performance management, customer service, telework policy?*
Business process expertise	*Does the person know how the organizational processes work e.g. how to get resources, the capabilities of an IT system?*
Methodology expertise	*Does the person have expertise in relevant methodologies e.g. consulting, facilitation, coaching, and project management?*
Recognized competence	*Is the person seen as credible and a good contributor?*
Independence	*Has the person a track record in thinking 'whole organization' and not 'my piece of it'.*
Availability	*Is the person willing and able to be available (and has this been cleared if necessary with his/her manager?)*
Authority	*Does the person have the authority and skills to make decisions, give advice, and recommend courses of action?*

The third step is to make sure the project team is calling on, and using the SME's skills. Not to be forgotten is the appropriate regular feedback, reward and recognition of SME work during the project duration. Here's an example of non-monetary reward – often the only type available - sent to the whole team, including SMEs, after a public event.

PMO Team,

You absolutely rocked the house! Your knowledge, presentation and command of difficult subject matter that other federal agencies are just barely scratching the surface on was clear! … ALL of you performed magnificently! …. Thank you, Thank you, Thank you! Your stars really shone brightly and you definitely represented our organization in the most excellent fashion!

KUDOS Team!

The Players

The SME is interacting with a range of players including:

- The program manager
- Tranche team members
- Other stakeholders (e.g. the SMEs line manager)
- Other SMEs
- Organizational employees
- Customers
- External organizations

Golden rules for SME interactions with the players:

- Focus on being credible. Provide good, useful, and usable information about the area of expertise tailored and appropriate for each of the players.
- Be original in your approach to your expertise and its application to the different players.
- Provide authoritative guidance in a way that doesn't come across as demanding or controlling.

The Outcome

Following the recipe above will result in expertise being put to good use. It will add value to the project deliverables and demonstrate a good return on investment in the SME role.

The Next Steps

- Think through the expertise you need for your project.
- Follow the recipe – making any suitable adaptations.
- Enjoy working with your SMEs.

www.ingramcontent.com/pod-product-compliance
Lightning Source LLC
Chambersburg PA
CBHW080622180526
45168CB00007B/3019